Star Light,
Star Bright

Hello and
Good-bye

Where
the Clouds
Go

 **HARCOURT BRACE JOVANOVICH, PUBLISHERS**
Orlando   San Diego   Chicago   Dallas

Where the Clouds Go

ODYSSEY An HBJ Literature Program
Second Edition

Sam Leaton Sebesta

Consultants

Elaine M. Aoki
Willard E. Bill
Sonya Blackman
Sylvia Engdahl

Myra Cohn Livingston
Daphne P. Muse
Sandra McCandless Simons
Barre Toelken

Acknowledgments

For permission to reprint copyrighted material, grateful acknowledgment is made to the following sources:

Atheneum Publishers: "Bedtime Stories" from *See My Lovely Poison Ivy* by Lilian Moore. Text copyright © 1975 by Lilian Moore. "Toaster Time" from *There Is No Rhyme for Silver* by Eve Merriam. Copyright © 1962 by Eve Merriam.
Atheneum Publishers and James Houston: "In summer the rains come" from *Songs of the Dream People: Chants and Images from the Indians and Eskimos of North America,* edited by James Houston (A Margaret K. McElderry Book). Copyright © 1972 by James Houston.
Curtis Brown, Ltd.: "Sunny Days" from *Kim's Place and Other Poems* by Lee Bennett Hopkins. Copyright © 1974 by Lee Bennett Hopkins.
William Collins Publishers, Inc.: From *The Secret Hiding Place* by Rainey Bennett. Copyright © 1960 by Rainey Bennett.
Delacorte Press / Seymour Lawrence: "Sebastian and the Bee" from *Journeys of Sebastian* by Fernando Krahn. Copyright © 1968 by Fernando Krahn. Copyright © 1968 by Dell Publishing Co., Inc.
Doubleday & Company, Inc.: "The Whale" by Buson from *Introduction to Haiku* by Harold G. Henderson. Copyright © 1958 by Harold G. Henderson.
E. P. Dutton, Inc.: Adapted from *The Gunniwolf* by Wilhelmina Harper. Copyright 1918, 1946 by Wilhelmina Harper.
E. P. Dutton and The Canadian Publishers, McClelland and Stewart Limited, Toronto: "Us Two" from *Now We Are Six* by A. A. Milne. Copyright 1927 by E. P. Dutton; renewed © 1955 by A. A. Milne.
James A. Emanuel: "A Small Discovery" by James A. Emanuel.
Grosset & Dunlap, Inc.: "The Horses" and the first and second verses of "Or Hounds to Follow on a Track" (Retitled: "I Wonder Where the Clouds Go?") from *The Sparrow Bush* by Elizabeth Coatsworth. Copyright © 1966 by Grosset & Dun-lap, Inc.
Harper & Row, Publishers, Inc.: "Okay everybody" from *Near the Window Tree: Poems and Notes* by Karla Kuskin. Copy-right © 1975 by Karla Kuskin. Text and specified illustra-tions from "Very Tall Mouse and Very Short Mouse" from *Mouse Tales* written and illustrated by Arnold Lobel. Copy-right © 1972 by Arnold Lobel.
Houghton Mifflin Company: First four lines from "I met a man with three eyes" and the first three lines from "I met a man on my way to town" in *I Met a Man* by John Ciardi. Copyright © 1961 by John Ciardi.
Macmillan Publishing Co., Inc.: "A Bear Went Over the Moun-tain" from *The Rooster Crows: A Book of American Rhymes and Jingles* by Maud and Miska Petersham. Copyright 1945 by Macmillan Publishing Co., Inc.; renewed © 1973 by Miska F. Petersham. "Who Has Seen the Wind?" from *Sing-Song* by Christina Rossetti. "The Little Turtle" from *Collected Poems* by Vachel Lindsay. Copyright 1920 by Macmillan Publishing Co., Inc.; renewed 1948 by Elizabeth C. Lindsay.

McGraw–Hill Book Company; "Message from a Caterpillar" from *Little Raccoon and Poems from the Woods.* Copyright © 1975 by Lilian Moore and Gioia Fiammenghi.

Penguin Books: "Where Go the Boats?" from *A Child's Garden of Verses* by Robert Louis Stevenson.

G. P. Putnam's Sons: "Hiding" from *Everything and Anything* by Dorothy Aldis. Copyright 1925, 1926, 1927; renewed 1953, 1954, 1955 by Dorothy Aldis.

Random House, Inc.: Specified excerpt (text only) from *He Bear, She Bear* by Stan and Jan Berenstain. Copyright © 1974 by Stan and Jan Berenstain.

Marian Reiner: "Tails" from *Whispers and Other Poems* by Myra Cohn Livingston. Copyright © 1958, 1978 by Myra Cohn Livingston.

Smithsonian Institution Press: "Send Us a Rainbow" in *Nootka and Quileute Music* by Frances Densmore from *Bureau of American Ethnology Bulletin 124,* page 285. United States Printing Office, Washington, DC, 1939.

The Viking Press: From *Gilberto and the Wind* by Marie Hall Ets. Copyright © 1963 by Marie Hall Ets. *Umbrella* by Taro Yashima. Copyright © 1958 by Taro Yashima.

Franklin Watts, Inc.: "Company" (Retitled: "Lunch for a Dinosaur") by Bobbi Katz, appearing in her book *Upside Down and Inside Out.* Copyright © 1973 by Bobbi Katz. Adapted from *Heather's Feathers* by Leatie Weiss. Copyright © 1976 by Leatie Weiss and Ellen Weiss.

Art Acknowledgments

Chuck Bowden: 20, 105, 124, 213 (adapted from photographs from the following sources: 20, Thomas Wommack, courtesy Harper & Row; 105, John Wheelock Freeman, courtesy Viking Penguin, Inc.; 124, courtesy Delacorte Press/Seymour Lawrence; 213, courtesy Viking Penguin, Inc.); Kinuko Craft: 19, 212; Sharon Harker: 164–165 top, 166–167 top, 186–187 top, 188–189 top; Mike Muir: 19; Jack Wallen: 147, 221, 222, 223, 224.

Cover: Tom Leonard.

Unit openers: Jane Teiko Oka.

Contents

10

1 Let's Go Together

VERY TALL MOUSE
and
VERY SHORT MOUSE

Story and pictures by Arnold Lobel

Once there was a very tall mouse
and a very short mouse
who were good friends.

When they met
Very Tall Mouse would say,
"Hello, Very Short Mouse."
And Very Short Mouse would say,
"Hello, Very Tall Mouse."

The two friends would often
take walks together.

As they walked along
Very Tall Mouse would say,
"Hello, birds."

And Very Short Mouse would say,
"Hello, bugs."

When they
passed by a garden,
Very Tall Mouse would say,
"Hello, flowers."
And Very Short Mouse
would say,
"Hello, roots."

When they passed by a house,
Very Tall Mouse would say,
"Hello, roof."
And Very Short Mouse
would say,
"Hello, cellar."

One day the two mice
were caught in a storm.

Very Tall Mouse said,
"Hello, raindrops."

And Very Short Mouse said,
"Hello, puddles."

They ran indoors to get dry.
"Hello, ceiling,"
said Very Tall Mouse.
"Hello, floor,"
said Very Short Mouse.

Soon the storm was over.
The two friends
ran to the window.

Very Tall Mouse held
Very Short Mouse up to see.

"Hello, rainbow!"
they both said together.

Questions

rainbow   birds   flowers   puddles   roots

1. What **two** things did
   Very Tall Mouse see?

2. What **two** things did
   Very Short Mouse see?

3. What did both friends see?

4. Which of these things could Very Tall Mouse
   see better than Very Short Mouse?

   a. ladder   **b.** bug   **c.** airplane

Activity   Draw Tall and Short Things

Draw something that is very tall.
Draw something that is very short.

About ARNOLD LOBEL

(To be read by the teacher)

Ever since he was seven
years old, Arnold Lobel
has been telling stories.
First he made up stories
for his friends at school,
then for his own children
when they were small. Now
Arnold Lobel is an author
and illustrator of books.
Many children enjoy his
stories and drawings.

Arnold Lobel says, "When I write my
stories, I always sit in the same chair. I do my
writing in the late afternoon. That is a good time
to think about frogs and toads and mice and crickets."

More Books by Arnold Lobel

Mouse Tales
Frog and Toad Together
Days with Frog and Toad
Owl at Home
Fables

Us Two

From the poem by A. A. Milne

Wherever I am, there's always Pooh,

There's always Pooh and Me.

Whatever I do, he wants to do,

"Where are you going today?" says Pooh:

"Well, that's very odd 'cause I was too.

Let's go together," says Pooh, says he.

"Let's go together," says Pooh.

CONNECTIONS

A Place to Live

This is a place where hippos live.

Other animals live here, too.

What animals do you see?

What plants do you see?

Animals must find food.

There is food here.

What food do you see?

This is a place where prairie dogs live.

Other animals live here, too.

What animals do you see?

What plants can you find?

What kind of food is here?

Questions

a.

b.

c.

d.

e.

f.

g.

h.

i.

1. Which animals live where the hippos live?

2. Which animals live where the prairie dogs live?

Activities

1. **Draw Animals and Living Places**

   Here are some places to live.

   Choose a place.

   Draw a picture of an animal that lives there.

forest   desert   pond

2. **Draw Animals and Food**

   Draw the hippo on a piece of paper.

   Then draw the kind of food it would eat.

step 1   step 2   step 3   step 4

The Secret Hiding Place

A story by Rainey Bennett

Pictures by Kinuko Craft

Little Hippo was the pet
of all the hippos.

Every morning the big hippos
waited for him to wake up.

Then they could take care of him.

"Sh," they whispered.

"Little Hippo is sleeping."

"Quiet all!" said Big Charles.

And every morning the big hippos
pushed and bumped each other.
They all wanted to be the first
to bring Little Hippo his breakfast.
Then they all settled down
to watch Little Hippo eat.

One morning Little Hippo felt cross.

"I don't want my breakfast,"
he said.

"I wish the hippos wouldn't watch
everything I do.

I wish I could be by myself
once in a while."

Big Charles put a cool leaf
on Little Hippo's head.

The leaf shaded him from the sun.

"Don't eat so fast,"
Big Charles said.

Big Charles took Little Hippo
for his morning walk.
All the hippos went along.
"We will protect you,"
said Big Charles.

But Little Hippo didn't want to be protected.

He wanted to go walking by himself.

What fun is a walk with nineteen hippos?

So without even saying "Excuse me, please," he dashed toward a bush.

"Stop, Little Hippo," Big Charles shouted.

"Birds nest there."

"Don't go in the tall grass.
Zebras hide there.
Do you want to catch stripes?"

Little Hippo stopped to look
at an ostrich with his head
in the sand.
"Come away, Little Hippo,"
Big Charles shouted.
"He thinks he's hiding."

Big Charles finally
caught up with Little Hippo.
He was hot and angry.
"That's a chameleon's house,"
Big Charles puffed.
"Come away right now!
When will you learn not to go
looking into secret places?"
Big Charles asked.

Everyone in the jungle
had a hiding place, it seemed.

Everyone but Little Hippo.

The leopard hid in shadowy places.

The tiger hid in the deep grass.

Birds hid in the trees.

Even the elephant was nearly hidden
by leaves as big as his ears.

"You're lucky," Little Hippo told the turtle and the snail.

"You carry your hiding places with you.

What's it like inside?"

"It's dark," said the turtle.

"It's dark," said the snail.

Little Hippo was
still cross at lunch.
    But after his nap
there was a big surprise.
    "We will play hide-and-seek,"
Big Charles said.
    "I will be IT."
    He leaned against a tree.
    Then he started to count
to five hundred by fives.

5
10
15
20
25
30
35
40

"Now!" Little Hippo whispered.

"Now I can find a hiding place
of my very own.

I'll hide in the river," he said.

"Little Hippo, Little Hippo,
hide with us."

"Oh, no!" Little Hippo said
to himself.

The lion laughed
when he saw Little Hippo
trying to crawl under a rock.

"Silly hippo," he said.
"That's no place to hide.
Follow me.
You can hide in my cave."

"Are we almost there?"
asked Little Hippo.

"Here we are," said the lion.
"Make yourself at home."
Then he went hunting for his dinner.

Little Hippo was all alone.

The dark cave was filled

with noises.

"I'm afraid," said Little Hippo.

"I don't want to be alone

this much."

Little Hippo was very frightened.

He ran out of the cave.

He ran for a long, long time.

Finally he sank to the ground
in a little heap.

"Whoof. I can't run any more,"
he said.

Just then the chameleon
put his head out of his house.

"Hello, Little Hippo," he said.

"What are you doing here?"

"I'm lost," said Little Hippo.

"You're lost?" said the chameleon.
"Follow me!"

He led Little Hippo
to the top of a small hill.
"Now, look, Little Hippo!"

And there right below him
was Big Charles.
He and all the other hippos
were looking for something.

"Little Hippo, come out,"
they called, pushing through the grass.
"Come out, come out,
wherever you are!" they shouted,
looking under rocks.
But not one of them thought
of looking up.

Little Hippo laughed and laughed.
"They'll never find me," he said.
"They don't see me
even though I'm right up here!"

"Home free! Home free!"
Little Hippo shouted.

He ran up to Big Charles.

All the big hippos were so glad
to see him.

They shouted and stamped their feet.

"Where did you hide, Little Hippo?

We looked everywhere,"
said Big Charles.

But Little Hippo didn't tell him.

He just smiled.

He knew that the big hippos

would always look everywhere but up.

And he never told anyone

about his secret hiding place

where he could be alone.

But not too alone.

Questions

| cave | tall grass | trees | top of a hill |

1.   Where does each one hide?

    a.   Zebras hide here.

    b.   The lion hides here.

    c.   Little Hippo hides here now.

    d.   Owls hide here.

2.   Which animal hides under a rock?

    a. worm   b. bird   c. bear

Activity   Draw Animals and Plants

Grasshoppers hide in grass.

Squirrels hide in trees.

Draw an animal hiding in a plant.

Write a sentence about your picture.

Hiding

A poem by Dorothy Aldis

I'm hiding, I'm hiding,
And no one knows where;
For all they can see is my
Toes and my hair.

Pictures by Lyle Miller

47

And I just heard my father
Say to my mother—
"But, darling, he must be
Somewhere or other;

Have you looked in the inkwell?"
And Mother said, "Where?"
"In the INK WELL," said Father. But
I was not there.

48

Then "Wait!" cried my mother—
"I think that I see
Him under the carpet." But
It was not me.

"Inside the mirror's
A pretty good place,"
Said Father and looked, but saw
Only his face.

49

"We've hunted," sighed Mother,
"As hard as we could,
And I AM so afraid that we've
Lost him for good."

Then I laughed out aloud
And I wiggled my toes
And Father said—"Look, dear,
I wonder if those

Toes could be Benny's.
There are ten of them. See?"
And they WERE so surprised to find
Out it was me!

50

Sunny Days

A poem by Lee Bennett Hopkins

Mile-long skyscrapers are my trees.
The subway's *whoosh,* my summer-breeze.

The hydrant is my swimming pool
Where all my friends keep real cool.

The city is the place to be.
The city is the place for me.

52

53

54

2 What a Surprise!

A Bear Went Over the Mountain

An American folk rhyme

A bear went over the mountain,
A bear went over the mountain,
A bear went over the mountain
To see what he could see.

The other side of the mountain,
The other side of the mountain,
The other side of the mountain
Was all that he could see!

Lunch for a Dinosaur

From the poem "Company" by Bobbi Katz

I'm fixing a lunch for a dinosaur.

Who knows when one might come by?

I'm pulling up all the weeds I can find.

I'm piling them high as the sky.

I'm fixing a lunch for a dinosaur.

I hope he will stop by soon.

Maybe he'll just walk down my street

And have some lunch at noon.

Pictures by Marie-Louise Gay

I Met a Man

From two riddles in verse by John Ciardi

I met a man on my way to town.

He was spinning up, he was spinning down.

He was twice as red as the nose of a clown.

(Mr. Yoyo.)

I met a man that was very wise.

He had no hands, but he had three eyes,

One green, one yellow, and one red.

He had nothing at all but eyes in his head.

(Mr. Traffic Light.)

Pictures by Sharon Harker

Heather's Feathers

A story by Leatie Weiss

Pictures by Ed Taber

The Tooth Fairy

Heather was the only bird
in her class.

She was the only one with feathers.
She was the only one with a beak.

Everyone thought Heather was great.
She could paint with her wings.
She could win every race.

And then everything changed.

It all began at snack time
one morning.

Robbie was having a hard time
eating his cookie.

"My front tooth is loose," he said.
"I can't wait for it to fall out."
He wiggled it with his paw.

"I lost two teeth already," said Patty.

"I lost four," said Amy.
"How many did you lose, Heather?"

"None," said Heather.
"You can't lose teeth
if you don't have teeth."

"No teeth?" they all asked.

"My beak can do everything
your teeth can do," said Heather.

"It can't get you a present
from the Tooth Fairy," said Robbie.
And they all started laughing
as if they knew a big secret.

Heather felt awful.

Then Robbie began shouting,
"My tooth fell out!"
What a fuss they all made.

"Put it under your pillow tonight,"
said the teacher.
"The Tooth Fairy will leave you
a present."

"So that's what the Tooth Fairy
does," thought Heather.
"What good is a beak?"

The next day Harold had lost
a tooth.

Then Freddy's tooth was loose,
too!

All they did was brag.

"I hate being a bird,"
thought Heather.

"I don't even believe
in the Tooth Fairy.

If she was a real fairy,
she'd help me grow a tooth!"

It was time for Show and Tell.

Heather had brought her toy bear.

But it was Harold's turn first.

He had 50 cents to show

from You Know Who.

"What did you bring today,
Heather?" asked the teacher.

Heather hid her bear.
"I forgot," she said,
and started to cry.

The teacher tried to cheer her up.
Then the bell rang.
It was time to go home.

A Surprise for Heather

It was Father Bird's turn
to drive home that day.
He saw Heather's sad face.

"Cheer up," said Father.
"I have a surprise for you.
I baked a cake
and you can help frost it."

69

"Who cares about cake?"
cried Heather.
"I want a surprise
from the Tooth Fairy.
I hate being a bird."

"Be proud you're a bird,"
said Father.
"We know a few tricks, too."

"What kind of tricks?"
asked Heather.

"You will see," said Father.
"Maybe someday soon."

"I don't believe it,"
cried Heather.
She gave her feathers a shake.

Feathers popped out!
She flapped her wings.
Out popped some more feathers.
"I'm getting bald!" she shouted.

"You're not getting bald,"
said Father.
"You are molting."

"Molting?" asked Heather.
"What's molting?"

"It's your surprise trick.
Your old feathers fall out.
Then bright new ones grow in,"
said Father.

POP

"I'm losing feathers,"
sang Heather.
"That's even better than
losing teeth!"

Heather began to dance.
Her feathers went flying.
She jumped into the pile.
"What do I do with them?"
she asked.

"Put them under your pillow," said Father.

POP

"Will the Tooth Fairy come?" asked Heather.

"No," Father laughed.
"But the Feather Fairy will."

POP

The Feather Fairy

That night, Heather thought
of the feathers under her pillow.
"I'm glad I'm a bird.
Feathers are much softer
to sleep on than teeth!"

The next day Heather got to school a bit late.

"I have something to show," she said.

"I lost lots of feathers because I am growing up.

And birds have a Feather Fairy!

Wait till you see the presents she left."

"I wish I had feathers," said Robbie.

"You can have some," said Heather.

"I got presents for everyone from the Feather Fairy."

"Aren't we lucky to have Heather
in our class!" said the teacher.

"Oh, yes!" shouted everyone.
"Heather's feathers are better
than the teeth we lost!"

Heather knew it was true.
And she popped out
a few loose feathers to prove it!

Questions

Heather was surprised.

Heather was happy.

Heather was sad.

1. What happened first?

2. What happened second?

3. What happened third?

4. Why did Heather lose her feathers?

   a. She was going bald.   **b.** She was sad.

   c. She was growing up.

5. "I am molting," said the _____.

   a. robin   **b.** bear   **c.** spider

6. What shows that Heather had a good father?

Activity   **Write About Proud Feelings**

Heather was proud when she lost her feathers.
Write a sentence to tell when you are proud.

Teeth, Teeth, Teeth

Birds have beaks, but they do not have teeth.

How do birds use their beaks?

Look at the pictures.

78

Pictures by Larry Frederick

Sharks have rows of teeth.

Old teeth fall out.

New ones grow in.

How do sharks use their teeth?

Look and see.

Elephants have six teeth.

Four teeth are in the elephant's mouth.

The elephant uses them to chew.

Two teeth are outside.

They are long.

They are strong.

They are called *tusks*.

How does the elephant use its tusks?

Dogs have many long teeth.

Dogs rub their teeth clean.

Bones help them clean their teeth.

What else helps them?

Look and see.

People have many small teeth.
What helps them care for
their teeth?
Look and see.
How do you care for your teeth?

The dentist can help take care
of your teeth.

Look at the picture.

How is the dentist helping?

How can you help the dentist?

Questions

Write the answer, **yes** or **no**.

1. Do elephants have rows of teeth?

2. Do sharks use their teeth to catch fish?

3. Are tusks long, strong teeth?

4. Do dogs rub their teeth clean?

5. Must birds take care of their teeth?

6. Must people take care of their teeth?

7. Do dentists help take care of people's teeth?

Activities

1. **Tell About Teeth Care**

   Show three ways that you can take care of your teeth.

2. **Make a Tooth Mobile**

   Draw a large tooth.

   Cut it out.

   Draw things that help you care for your teeth.

   Cut them out.

   Put them on your tooth mobile.

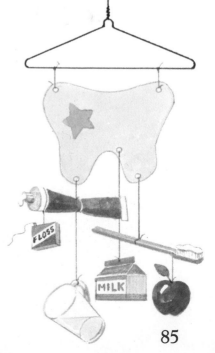

Toaster Time

A poem by Eve Merriam

Tick tick tick tick tick tick tick

Toast up a sandwich quick quick quick

Hamwich

Or jamwich

Lick lick lick!

Tick tick tick tick tick tick—stop!

POP!

Picture by Marie-Louise Gay

3 I Wonder

I Wonder Where the Clouds Go?

From a poem by Elizabeth Coatsworth

I wonder where the clouds go?
I wonder what the wind says?
I wonder what it is makes snow?
And how the birds get back?

90

I wonder how the flowers grow,
So many colors from one earth.
And how it is that feathers know
Which should be brown or red or black?

Pictures by Marie-Louise Gay

Rainbow Days

Two American Indian Poems

In summer the rains come,

The grass grows up,

and the deer has new horns.

—A Yaqui poem

You, whose day it is,
Make it beautiful.
Get out your rainbow colors,
So it will be beautiful.

—A Nootka poem

Who Has Seen the Wind?

A poem by Christina Rossetti

Who has seen the wind?
   Neither I nor you:
But when the leaves hang trembling
   The wind is passing through.

Who has seen the wind?
   Neither you nor I:
But when the trees bow down their heads
   The wind is passing by.

Picture by Christa Kieffer

About CHRISTINA ROSSETTI

(To be read by the teacher)

Christina Rossetti was born into a family of poets and artists in London, England. When she was twelve, she showed her own talent in a poem she wrote for her mother's birthday. Throughout her lifetime Christina Rossetti continued to write poetry and rhymes for children.

Many of Miss Rossetti's poems are about the special ways—the secrets—of nature. As you read her poems, you can share some of these secrets.

More Poems by Christina Rossetti

What Is Pink?
Sing-Song
Mix a Pancake

From

Gilberto
and the *Wind*

A story by Marie Hall Ets

Pictures by Mike Muir

I heard Wind whispering at the door.

"*You-ou-ou*,"

he whispers.

"*You-ou-ou-ou!*"

So I get my balloon,
and I run out to play.

At first Wind is gentle
and just floats my balloon
around in the air.

But then, with a jerk,
he grabs it away and carries it up
to the top of a tree.

"Wind! Oh, Wind!
Blow it back to me!
Please!"

But he won't.
He just laughs and whispers,

"*You-ou-ou-ou!*"

Wind likes my soap bubbles best of all.

He can't make the bubbles—
I have to do that.

But he carries them way up into the air for the sun to color.

Then he blows some back
and makes me laugh when they burst
on the back of my hand.

But then comes a day
when Wind is all tired out.

I whisper,
"Wind, oh, Wind!
Where are you?"

"*Sh-sh-sh-sh,*"

answers Wind,
and he stirs one dry leaf
to show where he is.

So I lie down beside him
and we both go to sleep—
under the willow tree.

Questions

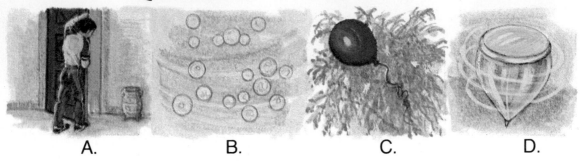

A. B. C. D.

1. Which picture shows when the wind *whispers*?

2. Which three pictures show when Gilberto likes the wind?

3. Which picture shows when Gilberto does not like the wind?

4. What can Gilberto play with if the wind stops blowing?

   a. sailboat **b.** kite **c.** top

Activity **Move with the Wind**

Move like a balloon in the wind.

About MARIE HALL ETS

(To be read by the teacher)

Marie Hall Ets has always felt a closeness to things in nature—animals, trees, flowers, grass, the clouds in the sky. That is why many of her books are about animals and other natural things. Her story *Gilberto and the Wind* tells of her feelings about the wind and about

Gilberto, a boy she met in California. Now that you have met Gilberto in the story *Gilberto and the Wind,* you might want to read about him again in the book *Bad Boy, Good Boy.*

More Books by Marie Hall Ets

Just Me
Another Day
Play with Me
In the Forest
Elephant in a Well

Sebastian

A picture story by Fernando Krahn

and the Bee

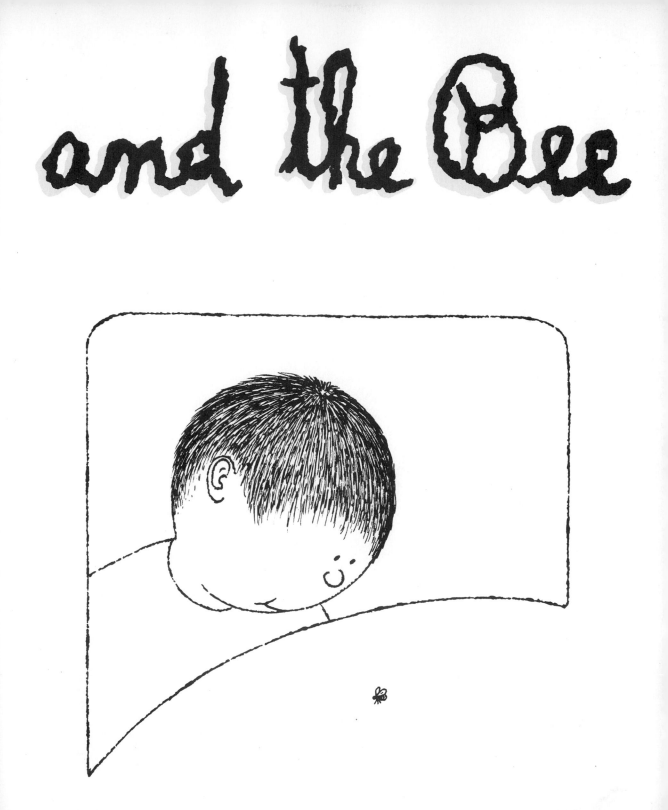

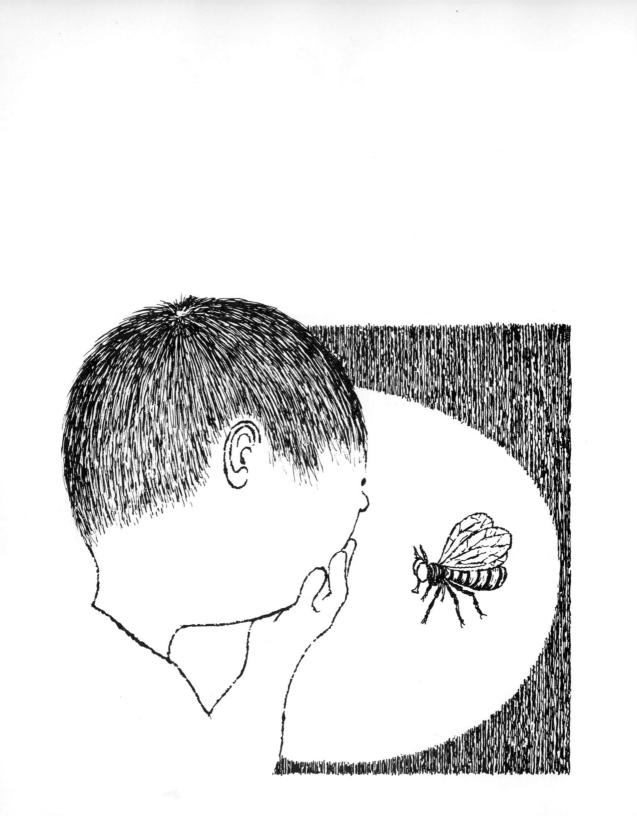

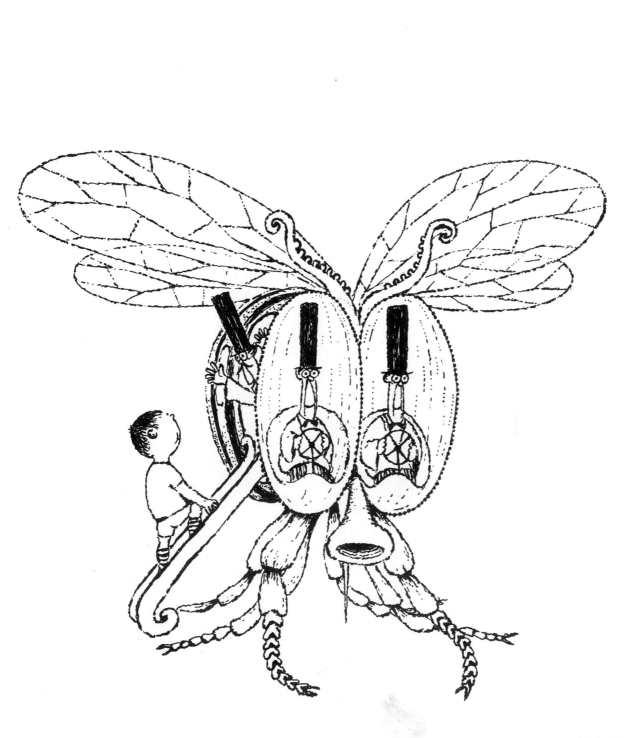

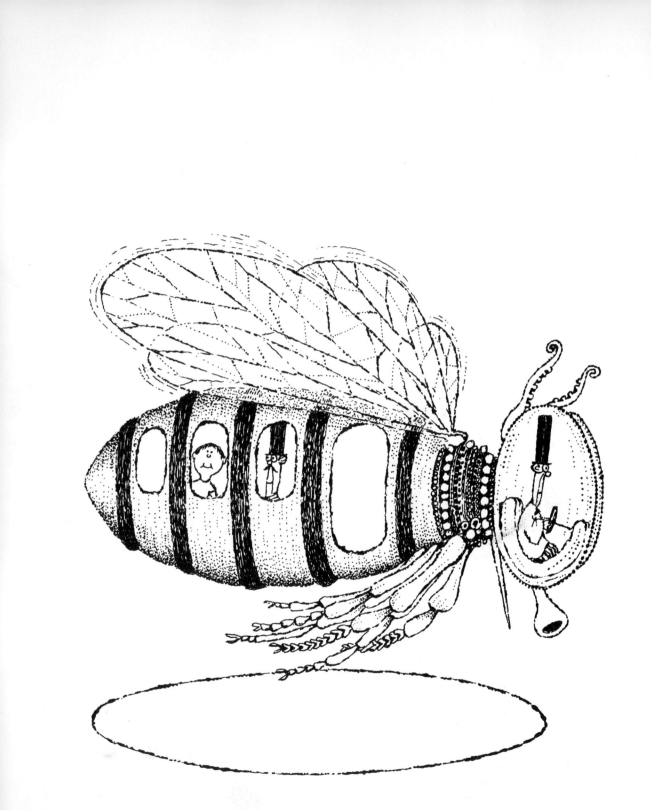

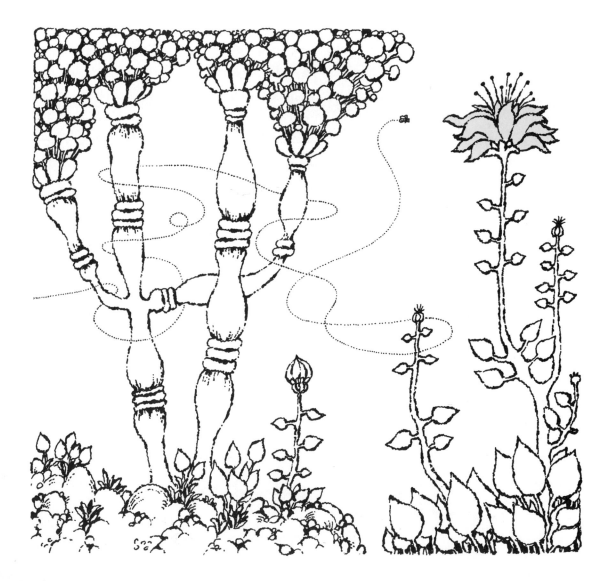

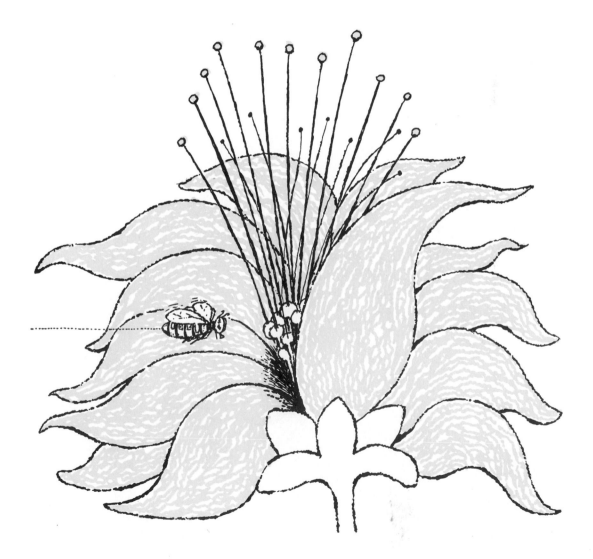

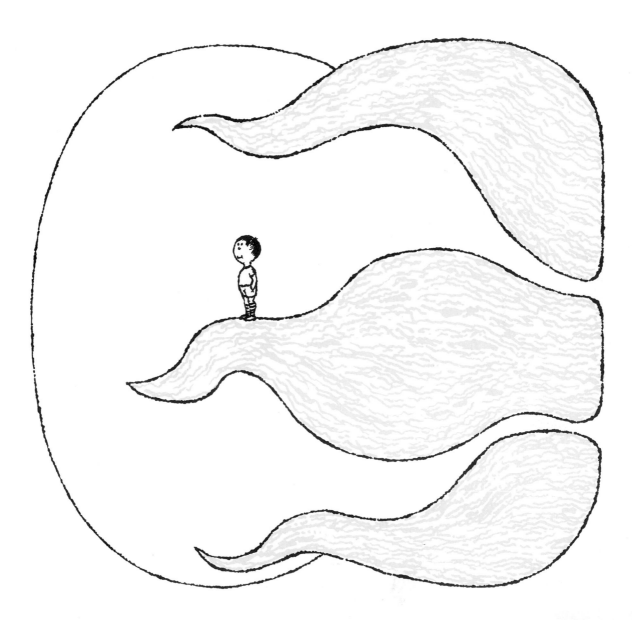

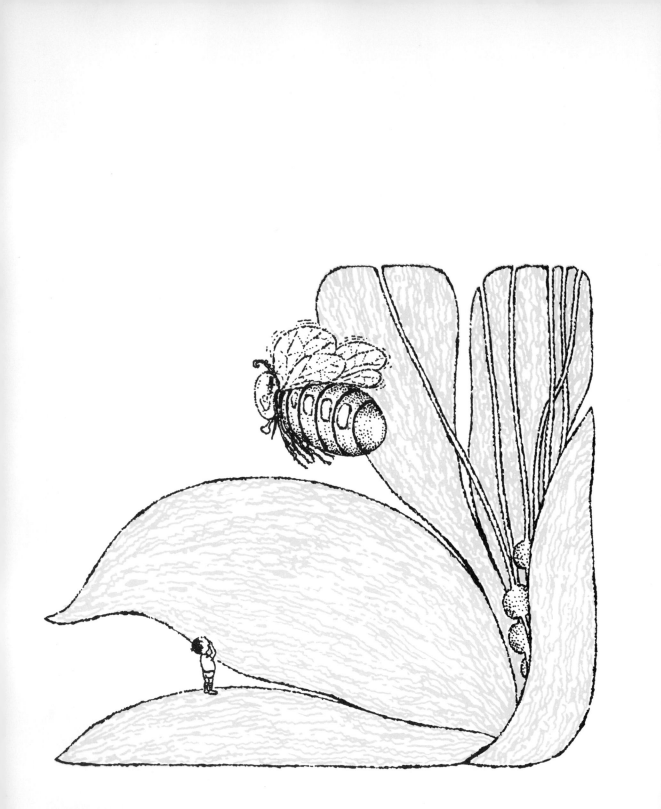

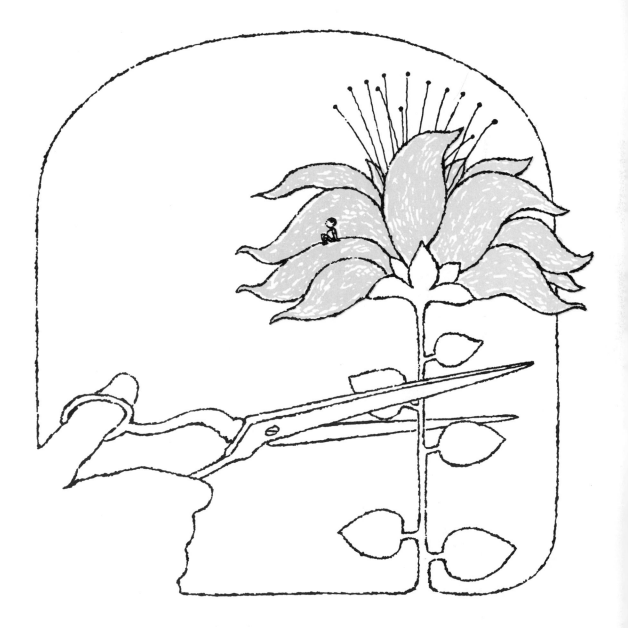

About FERNANDO KRAHN

(To be read by the teacher)

Do you like to draw pictures? Here is someone who has been drawing pictures since he was a child. He is Fernando Krahn, who drew the story *Sebastian and the Bee.*

Mr. Krahn, who was born in Chile, a country in South America, has said, "I can tell a story just through drawings, and that satisfies me greatly." Most of Mr. Krahn's books, which include several stories about the adventures of Sebastian, have no words. They are told through pictures alone.

More Books by Fernando Krahn

The Self-made Snowman
The Journeys of Sebastian
Sebastian and the Mushroom
The Mystery of the Giant Footprints
The Secret in the Dungeon

The Whale

A poem by Buson

A whale!
   Down it goes, and more and more
      up goes its tail!

Picture by Jane Teiko Oka

Tails

A poem by Myra Cohn Livingston

A dog's tail
    is short
And a cat's tail
    is long,
And a horse has a tail
    that he
    swishes along,
And a fish has a tail
    that can
    help him
    to swim,

Picture by Tony Kenyon

And a pig has a tail
that looks
curly on him.

All monkeys have tails
And all lions and whales.
There is
simply no
end
to
the
number
of
tails!

4 Far, Far Away

Where Go the Boats?

A poem by Robert Louis Stevenson

Dark brown is the river,
Golden is the sand.
It flows along forever,
With trees on either hand.

Green leaves a-floating,
Castles of the foam,
Boats of mine a-boating—
Where will all come home?

130

On goes the river,
And out past the mill,
Away down the valley,
Away down the hill.

Away down the river,
A hundred miles or more,
Other little children
Shall bring my boats ashore.

Picture by Christa Kieffer

THE GUNNIWOLF

A play adapted from an American folk tale
   retold by Wilhelmina Harper

Pictures by Jane Teiko Oka

Characters

Storyteller 1	Mother	Chorus
Storyteller 2	Little Girl	
Storyteller 3	Gunniwolf	

Storyteller 1: Once there was a little girl
who lived with her mother
in a house near a big woods.
Every day the mother would say,

Mother: Little Girl, you must never go
into the woods.
If you do, the Gunniwolf
might get you!

Storyteller 2: One day, the mother had to go away
for a little while.
Before she left, she said,

Mother: Now remember
to stay away from the woods.

Storyteller 3: And Little Girl said
she would remember.

Storyteller 1: When her mother was gone,
the girl saw
some beautiful white flowers.
They were growing at the very edge
of the woods.

Little Girl: Oh, what pretty flowers!
I'd like to pick just a few.

Storyteller 2: Little Girl went
to the edge of the woods
and began to pick the flowers.
As she did, she sang a little song.

Little Girl: Kum-kwa, khi-wa, kum-kwa, khi-wa.

Storyteller 3: Then she looked a little farther
into the woods.
Some beautiful pink flowers
were growing there.
Little Girl said,

Little Girl: Oh, I'll pick those, too!

Storyteller 1: So she went on into the woods
and began to pick the pink flowers.
As she did, she sang to herself.

Little Girl: Kum-kwa, khi-wa, kum-kwa, khi-wa.

Storyteller 2: Now the girl had her arms full
of pink and white flowers.
But way in the middle of the woods
she saw
some beautiful orange flowers.
So she said,

Little Girl: Oh, I'll pick just a few of those, too.
What pretty flowers I'll have
to show my mother!

Storyteller 3: Little Girl went still farther
into the woods and began to pick
the orange flowers.
As she did, she sang to herself.

Little Girl: Kum-kwa, khi-wa, kum-kwa, khi-wa.

Storyteller 1: SUDDENLY—
up rose the Gunniwolf and said,

Gunniwolf: Little Girl, why are you moving?

Storyteller 2: Little Girl was surprised,
but she answered,

Little Girl: I didn't move.

Gunniwolf: Then sing that sweet song again!

Storyteller 3: So she sang her song that went,

Little Girl: Kum-kwa, khi-wa, kum-kwa, khi-wa.

Storyteller 1: Soon the Gunniwolf nodded his head and fell fast asleep.

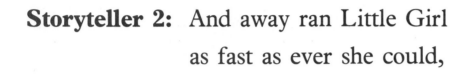

Storyteller 2: And away ran Little Girl
as fast as ever she could,

Chorus: Pit-pat, pit-pat, pit-pat, pit-pat—

Storyteller 3: Then the Gunniwolf woke up!
Away he ran after her,

Chorus: Hunker-cha, hunker-cha, hunker-cha—

Storyteller 1: When he caught up to her, he said,

Gunniwolf: Little Girl, why did you move?

Storyteller 2: And Little Girl answered,

Little Girl: I didn't move.

Gunniwolf: Then sing that sweet song again!

Storyteller 3: Softly Little Girl sang,

Little Girl: Kum-kwa, khi-wa, kum-kwa, khi-wa.

Storyteller 1: Soon the Gunniwolf nodded,
and nodded,
and went fast asleep.

Storyteller 2: Then AWAY ran Little Girl,

Chorus: Pit-pat, pit-pat, pit-pat, pit-pat—

Storyteller 3: She ran until she came
almost to the edge of the woods.

Chorus: Pit-pat, pit-pat, pit-pat, pit-pat—

Storyteller 1: She ran until she got away
out of the woods.

Chorus: Pit-pat, pitty-pat, pit-pat, pitty-pat—

Storyteller 2: She ran until she reached
her very own door.

Storyteller 3: From that day on,
Little Girl has never, NEVER gone
into the woods again.

Questions

Who am I?

Little Girl

Mother

Gunniwolf

1. I said, "Never go into the woods."

2. I said, "Oh, what pretty flowers!"

3. I said, "Sing that sweet song again!"

4. I sang, "Kum-kwa, khi-wa."

5. I ran "hunker-cha, hunker-cha."

6. The next day, Little Girl wanted some flowers. How did she get the flowers?

Activity   **Draw a Scary Animal**

Make up your own scary animal.

Draw its picture.

Cut it out.

Write made-up words to tell how it moves.

The Horses

A poem by Elizabeth Coatsworth

Red horse,
Roan horse,
Black horse,
And white,
Feeding all together in the green
    summer light.

White horse,
Black horse,
Spotted horse,
And gray,
I wish that I were off with you,
    far, far away!

Picture by Kazuhiko Sano

5 Tell Me a Story

Bedtime Stories

A poem by Lilian Moore

"Tell me a story,"
Says Witch's Child.

"About the Beast
So fierce and wild.

About a Ghost
That shrieks and groans.

A Skeleton
That rattles bones.

About a Monster
Crawly-creepy.

Something nice
To make me sleepy."

A Small Discovery

A poem by James A. Emanuel

Father,
Where do giants go to cry?

To the hills
Behind the thunder?
Or to the waterfall?
I wonder.

(Giants cry.
I know they do.
Do they wait
Till nighttime too?)

Pictures by Kinuko Craft

The Three Billy Goats Gruff

A Norwegian folk tale collected by P. C. Asbjörnsen
and Jörgen E. Moe

Pictures by Willi Baum

Once on a time there were
three Billy Goats who were to go up
to the hillside to make themselves fat.
And the name of all three
was "Gruff."

On the way up was a bridge
over a river they had to cross.
And under the bridge
lived a great ugly Troll,
with eyes as big as saucers
and a nose as long as a poker.

155

So first of all came
the youngest Billy Goat Gruff
to cross the bridge.

"*Trip, trap! Trip, trap!*"
went the bridge.

"*Who's that* tripping
over my bridge?" roared the Troll.

"Oh! It is only I,
the tiniest Billy Goat Gruff.
I'm going up to the hillside
to make myself fat," said the Billy Goat,
with such a small voice.

"Now, I'm going to gobble you up," said the Troll.

"Oh, no! Please don't take me. I'm too little," said the Billy Goat. "Wait a bit 'til the second Billy Goat Gruff comes. He's much bigger."

"Well! Be off with you," said the Troll.

A little while after
the second Billy Goat Gruff
came to cross the bridge.

"TRIP, TRAP! TRIP, TRAP!
TRIP, TRAP!" went the bridge.

"WHO'S THAT tripping
over my bridge?" roared the Troll.

"Oh! It's the second
Billy Goat Gruff.
I'm going up to the hillside
to make myself fat," said the Billy Goat,
with a voice that was not so small.

"Now, I'm going to gobble you up,"
said the Troll.

"Oh, no! Don't take me.
Wait a little 'til
the big Billy Goat Gruff comes.
He's much bigger."

"Very well! Be off with you,"
said the Troll.

159

Just then up came
the big Billy Goat Gruff.

"TRIP, TRAP! TRIP, TRAP!
TRIP, TRAP! TRIP, TRAP!"
went the bridge.

"WHO'S THAT tramping
over my bridge?" roared the Troll.

"IT'S I!
THE BIG BILLY GOAT GRUFF,"
said the Billy Goat,
who had a big, ugly voice of his own.

"Now, I'm going
to gobble you up,"
roared the Troll.

"Well, come along!"
said the big Billy Goat Gruff.
"I've got two big spears
to fight you with."

So he flew at the Troll
and tossed him out into the river.

Then he went up to the hillside.
There the three Billy Goats
got so fat that they were not able
to walk home again.

And if the fat hasn't fallen off them,
why they're still fat.

And so—

"Snip, snap, snout,
This tale's told out."

Questions

bridge

Who am I?

goat

Troll

1. I gobble up goats.

2. I go "trip, trap, trip, trap."

3. I eat grass.

4. What did the Troll learn?

    a. Hide in the river.

    b. Stand on the bridge.

    c. Do not bother goats.

Activity   Write a Story Riddle

Read this riddle.

> I have eyes as big as saucers.
> I have a nose as long as a poker.
> Who am I?

Make up a new "Who am I?"

Learn About

Stories

One, Two, Three—Surprise!

Some stories tell about things
that happen one, two, three times.
Then there is a surprise ending!

Here is a story you know.
Draw or tell the end of the story.

The Three Billy Goats Gruff

What is the surprise?

Now tell your own story.

Maggie reached into her magic hat one, two, three times.

Tell or draw what happens the last time.

Make your ending a surprise.

Maggie's Magic Hat

What is the surprise?

The Little Turtle

A poem by Vachel Lindsay

Pictures by Larry Mikec

There was a little turtle.

He lived in a box.

He swam in a puddle.

He climbed on the rocks.

He snapped at a mosquito.

He snapped at a flea.

He snapped at a minnow.

And he snapped at me.

He caught the mosquito.
He caught the flea.
He caught the minnow.
But he didn't catch me.

6 I'm Growing

Okay Everybody!

A poem by Karla Kuskin

Okay everybody, listen to this:

I am tired of being smaller

Than you

And them

And him

And trees and buildings.

So watch out

All you gorillas and adults

Beginning tomorrow morning

Boy

Am I going to be taller.

Message from a Caterpillar

A poem by Lilian Moore

Picture by Richard Katz

Don't shake this
bough.
Don't try
to wake me
now.

In this cocoon
I've work to
do.
Inside this silk
I'm changing
things.

I'm worm-like now
but in this
dark
I'm growing
wings.

177

CONNECTIONS

Penguins

When spring comes, the penguins
make a nest.
They pile up little rocks.

Pictures by Tom Dunnington

The penguins' nest is ready!
Mother penguin lays an egg.
She keeps it warm.
Father penguin helps.

The egg cracks!
Out comes the baby penguin.
The baby needs food.
Mother goes to catch fish.
Father keeps the baby warm.

Mother comes back with food.

She feeds the baby.

Now it is Father's turn.

Mother will keep the baby warm.

Father will catch fish.

The baby penguins are growing!

They are almost as big as their
mothers and fathers.

Now they can stay alone.

Their mothers and fathers go
fishing.

The babies crowd together to keep
warm.

Penguins have wings, but they do not fly.

They walk on the ice.

Wings keep the penguins from falling or slipping.

Look at the picture.

How do penguins use their wings in the water?

Now the baby penguins are grown.
They do not look like babies.
How have they changed?

Someday they will make nests.
There will be more baby penguins!

A.

B.

Questions

C.

1. What happened first?
2. What happened next?
3. What happened then?
4. What happened last?

D.

Activity   Draw a Growing Pattern

Once you were a baby.

Someday you will be grown up.

Draw three pictures.

Show how you looked when you were a baby.

Show how you look now.

Show how you might look when you are grown up.

Learn About

Stories

Story Sounds

Some words in stories tell how things sound.

When the Wind blows,
the sound is
"you-ou-ou."

When the Gunniwolf runs,
the sound is
"hunker-cha, hunker-cha."

When the Billy Goats Gruff walk
over the bridge,
the sound is
"trip, trap, trip, trap."

186

Pictures by Ed Taber

Get ready to read a story about us.

These things are in the story.

Make up words to tell how each thing sounds.

Say your sounds for these things
as you read the story
on the next page.

The Hare and the Tortoise

The hare and the tortoise
had a race.

The hare started.

The tortoise started, too.

The hare swam across a lake.

The tortoise swam across the lake, too.

The hare went
through a pile of leaves.

The tortoise went
through the pile of leaves, too.

The hare took a nap.

The tortoise kept going.

The hare woke up and ran.

But the tortoise won the race!

UMBRELLA

Adapted from the story by Taro Yashima

Pictures by Taro Yashima

Momo (Peach)

Momo is the name of a little girl
who was born in New York.
The word _Momo_ means
"the peach" in Japan
where her father and mother
used to live.

On her third birthday
Momo was given two presents—
red rubber boots and an umbrella!
They pleased her so much
that she even woke up that night
to take another look at them.

Natsu (Summer)

But it was still summer,
and the sun was bright.

Every morning
Momo asked her mother,
"Why doesn't the rain fall?"
The answer was always the same,
"Wait, wait. It will come."

195

Ame (Rain)

It was many, many days later
that finally the rain fell.
Momo was wakened
by her mother calling,
"Get up! Get up!
What a surprise for you!"

197

Momo did not stop to wash her face.
She even pulled the boots
onto her bare feet—
she was so excited.

The sidewalk was all wet and new.
Raindrops were jumping all over,
like tiny people dancing.

The street was crowded and noisy.
But she said to herself,
"I must walk straight,
like a grown-up."

On the umbrella,
raindrops made a wonderful music
she never had heard before—

    Bon polo

    bon polo

    ponpolo ponpolo

    ponpolo ponpolo

    bolo bolo ponpolo

    bolo bolo ponpolo

    boto boto ponpolo

    boto boto ponpolo

The rain did not stop all day long.
Momo watched it at times
while she was playing the games
at the nursery school.

She did not forget her umbrella
when her father came
to take her home.
She used to forget
other things so easily—
but not her umbrella.

The street was crowded and noisy.

But she said to herself,

"I must walk straight,

like a grown-up!"

On her umbrella,

the raindrops made the wonderful music—

> *Bon polo*

> *bon polo*

> *ponpolo ponpolo*

> *ponpolo ponpolo*

> *bolo bolo ponpolo*

> *bolo bolo ponpolo*

> *boto boto ponpolo*

> *boto boto ponpolo*

all the way home.

Questions

1. Momo asked, ''Why doesn't the rain fall?'' Why did she ask that?

2. What were the raindrops like?

   a. Tiny pebbles falling
   b. Tiny people dancing

3. How did Momo show she was growing up?

   a. She watched the rain.
   b. She did not forget her umbrella.
   c. She went to school by herself.

Activity   Write Sound Words

Momo's boots made wonderful music when she walked in the rain.

Make up words for that wonderful music.

Write the words you have made up.

About TARO YASHIMA

(To be read by the teacher)

As soon as Taro Yashima
(TAH·roh YAH·shee·mah)
gets an idea for a story, he
writes it on a piece of
paper. Then he puts the
paper in an envelope
until he is ready to write
the story.

Many of his stories
began because his
daughter, Momo, wanted to
know what he did as a boy in Japan.
A few of Taro Yashima's stories are about
Momo. He wrote one of his favorite stories,
Umbrella, as a gift for Momo's eighth birthday.

More Books by Taro Yashima

Crow Boy
Seashore Story
Momo's Kitten
Village Tree
Youngest One

He Bear, She Bear

Adapted from the story in verse by Stan and Jan Berenstain

Pictures by Tony Kenyon

There are many things to be.
Come on, He Bear, follow me!

We could fix a clock,
Paint a door,
Build a house,
Have a store.

We could be doctors
And make people well.
We could teach kids
How to add and spell.

I may build bridges,
I may climb poles,
I may race cars,
I may dig holes.

I could be a magician,
I could go on TV.
I could study the fish
Who live in the sea.

We'll jump and dig and build and fly. . .

There is *nothing* that we cannot try.

We can do all these things, you see,

Whether we are he OR she.

Key Words

Sounds and Letters

/sk/
skeleton page 152

/sch/
school page 204

/sl/
sleep page 103

/sm/
smile page 45

/sn/
snail page 36

/sp/
spotted page 148

/st/
storm page 17

/str/
stripes page 33

/sw/
swim page 126

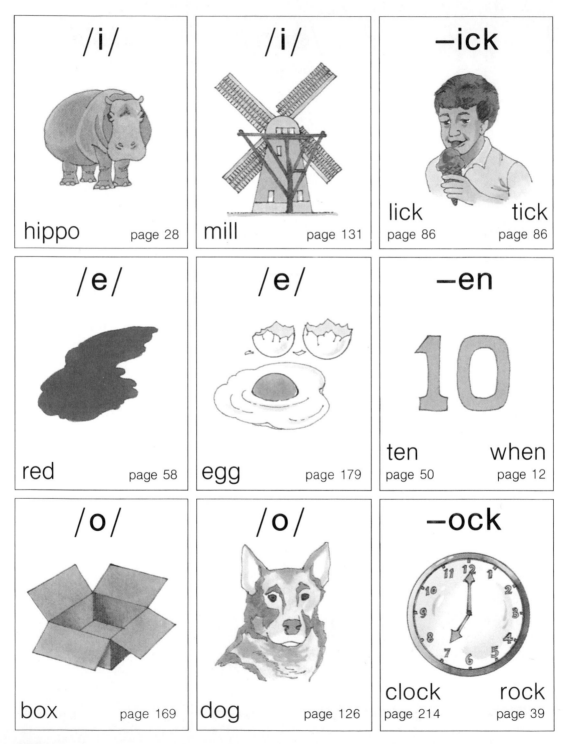

/i/

hippo — page 28

/i/

mill — page 131

–ick

lick — page 86

tick — page 86

/e/

red — page 58

/e/

egg — page 179

–en

ten — page 50

when — page 12

/o/

box — page 169

/o/

dog — page 126

–ock

clock — page 214

rock — page 39

/ā/

cave
page 39

/ā/

whale
page 125

−ake

lake
page 188

wake
page 28

/ī/

mice
page 16

/ī/

white
page 148

−ive

5

five
page 37

drive
page 69

/ō/

bones
page 81

/ō/

nose
page 58

−ole

pole
page 216

hole
page 216

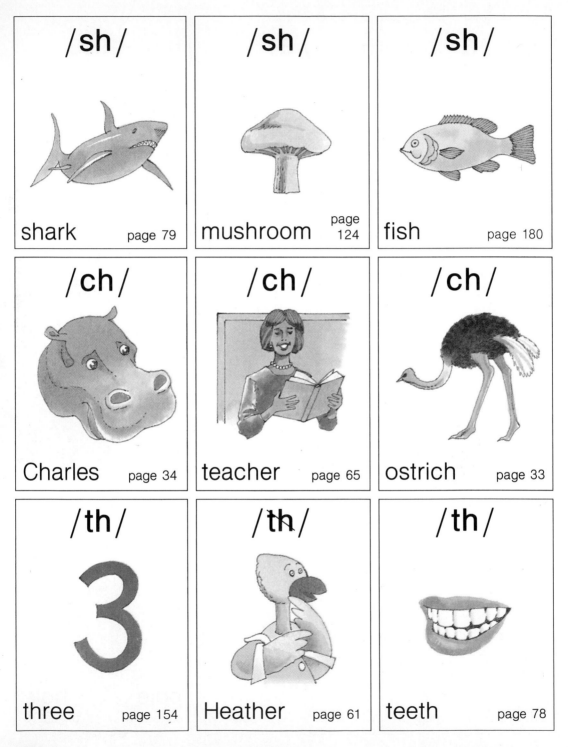

/sh/

shark   page 79

/sh/

mushroom   page 124

/sh/

fish   page 180

/ch/

Charles   page 34

/ch/

teacher   page 65

/ch/

ostrich   page 33

/th/

three   page 154

/th/

Heather   page 61

/th/

teeth   page 78